Albatros Fighter Aircraft of WW1

Dave Douglass

Introduction

As an illustrator, aircraft nut, plastic modeller, and Photoshop professional, I created the artwork in this book with all of these disciplines combined with a lifelong obsession with WW1 German aircraft and markings, more specifically, the Albatros D series biplane fighter. The Albatros is perfect for the genre of aircraft profile art, not only because of its beautifully finished fuselage of varnished plywood, painted metal fittings, pleasing shape, and variety of factory finishes, but also because of its myriad *Jagdstaffe* and personal markings. The D.I and D.II started their operational careers at the same time as the formation of the *Jagdstaffen*. These *Jagdstaffen* adopted individual markings to distinguish themselves at the same time that more pilots were adopting personal markings for the same purpose. This produced a seemingly endless combination of colours and patterns.

The profiles in this book were started in February of 2011. The variants of the airframe were not completed until May of 2013. They were created on an Apple Mac Pro using Photoshop. The methods employed are very similar to those used to create traditional technical airbrush art using frisket masks. Very few "automatic" Photoshop techniques were used and no 3D modeling was employed. They are strictly interpretations. Black and white photos were used as a primary source along with published research and correspondence with WW1 markings experts. The colours used are based on research accumulated by many authors and enthusiasts, first-hand accounts, interviews with veterans, combat reports, and surviving relics. The study of colour used on aircraft from the Great War can never be exact, as has been noted by many authors who have written about this subject. The paints, dopes, and pigments on surviving relics are subject to natural weathering as well as changes to carriers and protective varnishes used. First-hand accounts can be equally open to interpretation. Seemingly specific colour references such as "Brunswick green" or "Venetian red" can provide a large range of variations. Familiarity with artist colour theory can only help point out the many possibilities rather than find an exact match.

Photographic practices of the time also make any conclusions difficult. Orthochromatic film can make light colours seems dark and dark colours seem the opposite. Blue and yellow are especially troubling. Lighting conditions make this even more difficult. Photographs of the same plane taken at different times can be very enlightening. Lastly, many of the surviving photos are copies of older photographs which themselves may have faded or otherwise changed over the years. It can never be assumed that any print is made from a negative in perfect archival conditions.

I would like to thank Jim Miller and Greg VanWyngarden for graciously sharing their exceptional and endless knowledge and encouragement, Arjan "Ser
Logt and everyone at simmerspaintshop.com for providing such a wonderf
all aircraft artists, theaerodrome forum for accumulating so much research i
Albatros Publications for all of their groundbreaking books and magazine
Brown and Ugo Crisponi for their online support, Ryan Toews for helpi
out, and especially the late Dan-San Abbott for always being around when

My work can be found online at:
www.bravobravoaviation.com
www.aviationgraphic.com

1

Albatros D.I 391/16

Ltn. Karl Heinrich Otto Büttner, Jasta 2, Lagnicourt, November 1916

Red-brown and green surfaces, light blue undersides. Fuselage partially over-painted in red-brown.

Albatros D.I 391/16

Captured Aircraft No. G.1

Ltn. Büttner of Jasta 2 was shot down by BE2c crew Capt. Parker and Ltn. Hervey of No. 8 Squadron R.F.C. on 16 November 1916. After 391/16 was captured it was extensively tested and studied by the R.F.C. in Britain. Possibly painted in PC-10.

3

Albatros D.I 457/16

Jasta 6, Ugny, November 1916

Finished in dark green/light green/red-brown upper surfaces, light blue undersides.

Albatros D.I

Prince Friedrich Karl of Prussia, Jasta "Boelcke", Eswars, March 1917

Prince Friedrich Karl was forced to land behind enemy lines in this plane on 21 March 1917.
Overall light green with black and white markings.

5

Albatros D.II (O.A.W.) 910/16

Ltn. Max Böhme, Jasta 5, Gonnelieu, March 1917

Overall O.A.W. camouflage of dark green/light green/red-brown upper surfaces and light blue undersides.

Albatros D.II AL910, ex D.910/16

Captured by British forces on 4 March 1917

Repainted in French overall aluminum dope and tested at Villacoublay, March 1917.

7

Albatros D.II (O.A.W.) 933/16

Vzfw. Jakob Wolff, Jasta 17, Metz, February 1917

Finished in O.A.W. camouflage of dark green/light green/red-brown upper surfaces and light blue undersides.

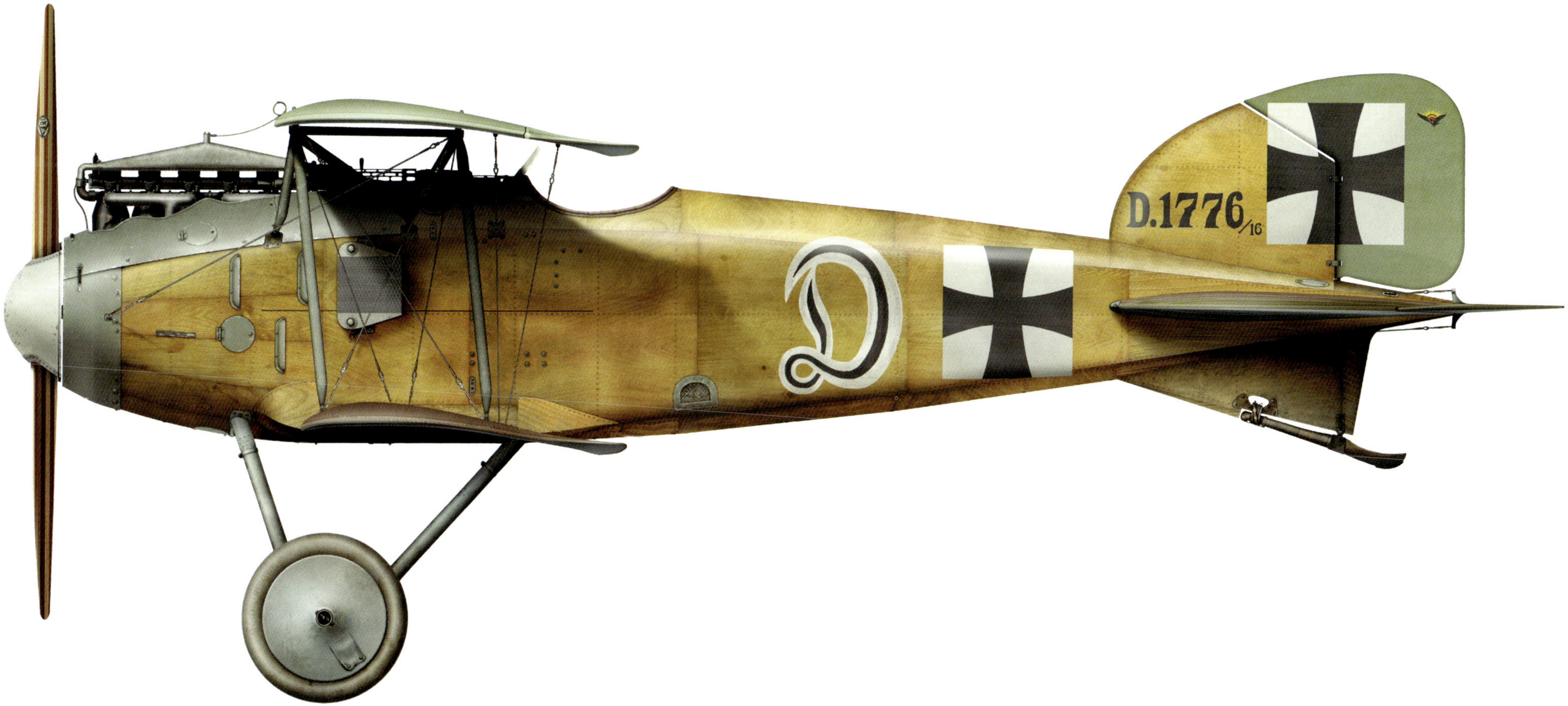

Albatros D.II 1776/16/16

Vzfw. Walter Dittrich, Jasta 1, Vivaise, April 1917

Finished in dark green/light green/red-brown upper surfaces and light blue undersides.

Albatros D.II (L.V.G.)

Oblt. Paul Kremer, Jasta 16b, Ensisheim, March 1917

Surfaces finished in L.V.G. camouflage of dark green/light green/red-brown upper surfaces and light blue undersides.

Albatros D.III D.2096/16

Ltn. Friedrich-Wilhelm Wichard, Jasta 24, Annelles, April 1917

Surfaces finished in dark green/light green/red-brown upper surfaces and light blue undersides.

11

Albatros D.III D.2096/16

Ltn. Friedrich-Wilhelm Wichard, Jasta 24, Captured by French forces, 21 April 1917

Sent to McCook Field in Dayton, Ohio, U.S.A. And exhibited at the New York City Aero Show, February 1919.

Albatros D.III 2183/16

Lt. Hans Fritzsche, Jasta 29, 1917.

Finished in dark green/light green/red-brown upper surfaces and light blue undersides.

13

Albatros D.III “Le-Petit-Rouge”

Rttm. Manfred von Richthofen, Jasta 11, Roucourt, April 1917

Red personal and Jasta markings.
Upper surface of top wing in green/light green/red-brown with light blue undersides. Lower wing with red upper surfaces and light blue undersides.

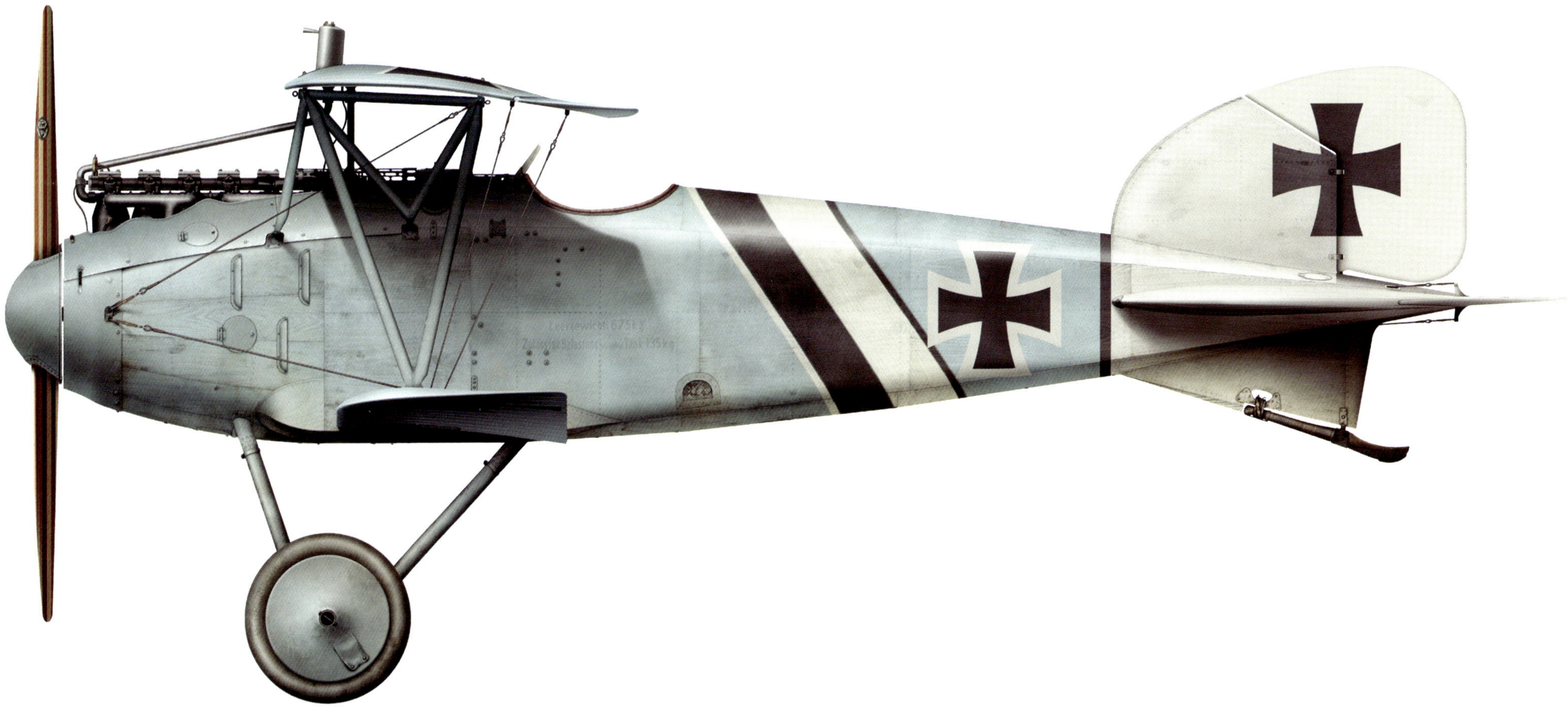

Albatros D.III

Ltn. Hermann Frommherz, Jasta 2 "Boelcke", Pronville, April 1917

Overall light blue with black and white personal and Jasta markings.

15

Albatros D.III D.2219/16

Ltn. Gerhard Bassenge, Jasta 2 "Boelcke", Pronville, June 1917

Black and white personal and Jasta markings.
Finished in dark green/light green/red-brown upper surfaces and light blue undersides.

Albatros D.III (O.A.W.)

Ltn. Kurt Wusthoff, Jasta 4, Summer 1917

Black and white personal and Jasta markings.
Finished in green/mauve upper surfaces and light blue lower undersides.

17

Albatros D.III

Ltn. Werner Voss, Jasta 5, Krefeld, June 1917

Finished in dark green/light green/red-brown upper surfaces and light blue undersides.
***Final version featuring replacement wing with offset radiator.**

Albatros D.III (O.A.W.)

Jasta 39, Veldes, Slovenia, October 1917

Overall red (or black?) and white stripes, light blue lower fabric surfaces.

19

Albatros D.III (O.A.W.) D.2576/17

Jasta 46, Ascq, February 1918

Black and white personal markings.
Upper surfaces and rudder in five-colour lower/light printed fabric, light blue undersides.

Albatros D.V Prototype, D.388/16

Johannisthal, September 1916

All fabric surfaces in five-colour upper and lower printed fabric.
Non-fabric surfaces painted in five-colour upper and lower camouflage colours.

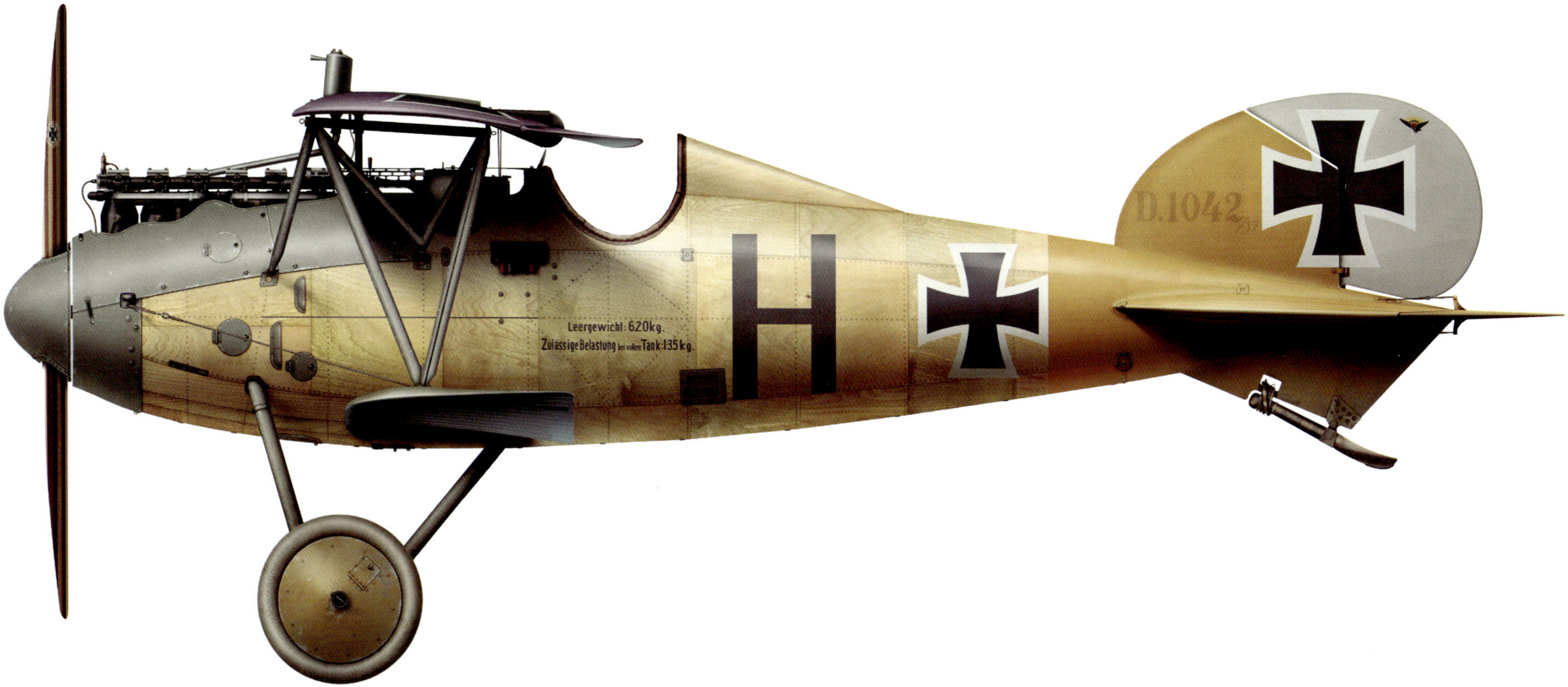

Albatros D.V D.1045/17
Jastaschule 1, Valenciennes 1917

Yellow and black personal markings.
Wings in factory green/mauve upper, light blue lower camouflage. Pale grey-green metal cowling and fixtures.

Albatros D.V D.1119/17

Oblt. Ernst Freiherr von Althaus, Jasta 10, Heule, July 1917

Yellow fuselage with black and white personal markings.
Wings in factory green/mauve upper, light blue lower camouflage.

23

Albatros D.V 2034/17

Ltn. Eduard Ritter von Schleich, Jasta 21, Chassogne Ferme, May 1918

Jasta markings in light blue/green.
Wings in factory green/mauve upper, light blue lower camouflage.

Albatros D.V

Ltn. Kurt Monnington, Jasta 18, Houplin, November 1917

Black and white personal markings. Upper wing in factory green/mauve upper, light blue lower camouflage.
Lower wing in five colour upper and lower printed fabric. Rudder and stabilizer in light five-colour printed fabric.

Albatros D.V D.2059/17

Rttm. Manfred von Richthofen, JG. I, Marckebeke, August 1917

Red personal and Jasta markings.
Wings in factory green/mauve upper, light blue lower camouflage.

Albatros D.V

Ltn.d.R. Ernst Udet, Jasta 37, Wynghene, Autumn 1917

Painted in overall silver-grey with black metal cowling and fixtures.
Jasta 37 tail markings in white and black.

27

Albatros D.V

Possibly flown by Ltn. Fritz Kempf, Jasta "Boelcke", Varsenaere, September 1917

Black and white personal and Jasta markings.
Wings in factory green/mauve upper, light blue lower camouflage.

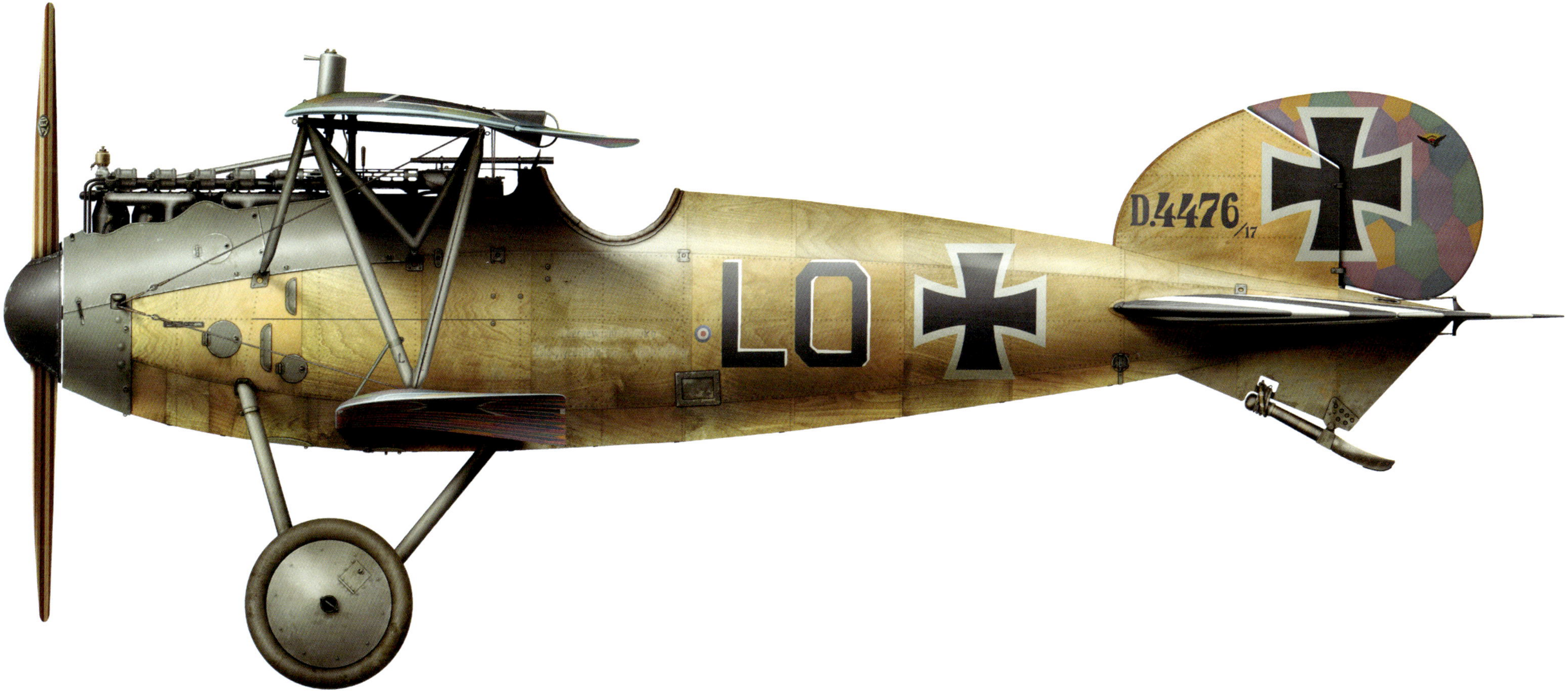

Albatros D.V D.4476/16

Ltn.d.R. Ernst Udet, Jasta 37, Phalempin, September 1917

Jasta 37 and personal markings in white and black.
All surfaces in five-colour printed fabric. Pale grey-green metal cowling and fixtures.

29

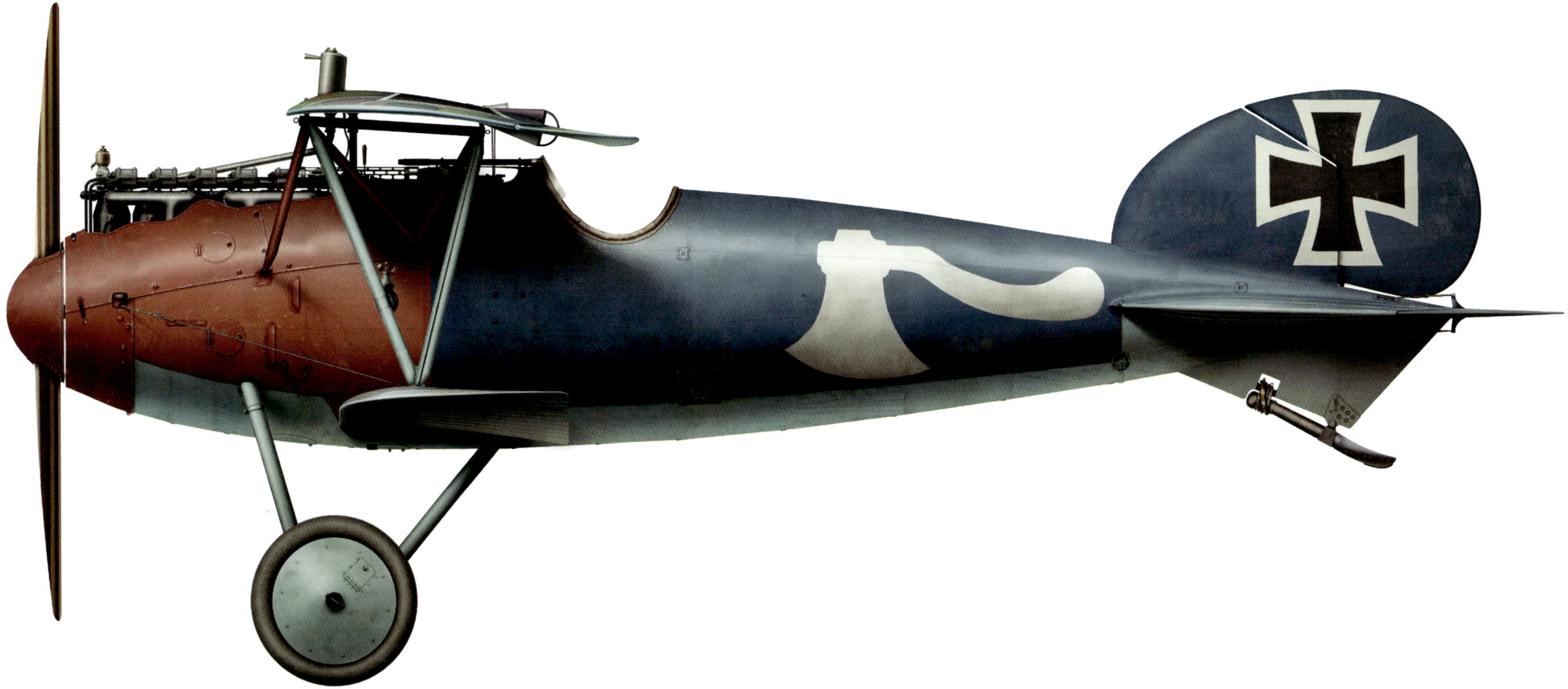

Albatros D.V D.4594/17

Ltn.d.R. Paul Strahle, Jasta 18, Harlebecke, November 1917

Fuselage in red/blue/light blue Jasta markings.
Wings in five-colour printed fabric upper surfaces and light blue lower surfaces.

Albatros D.V 2065/17

Oblt. Richard Flasher, later flown by Ltn. Hans Joachim von Hippel

Jasta 5, Boistrancourt, February 1918

Jasta markings, grey/red/green.
Wings in factory green/mauve upper, light blue lower camouflage.

31

Albatros D.V

Oblt. Richard Flasher, Jasta 5, Cappy, April 1918

Jasta markings, grey/red/green.
Wings in factory green/mauve upper, light blue lower camouflage.

Albatros D.Va

Ltn Hans Bohning, Jasta 76b, Villiers-le-Sec, December 1917

Fuselage in white/light blue/dark blue Jasta and personal markings.
Fabric surfaces in five-colour printed fabric.

Albatros D.Va

Ernst Udet, Wynghene, Jasta 37, Winter 1917

Black fuselage with black and white personal and Jasta markings.
Fabric surfaces in five-color printed fabric.

Albatros D.Va

Ltn. Kurt Monnington, Jasta 18, Lomme, May 1918

Upper surfaces and fuselage in red and white Jasta markings.
Lower fabric surfaces in five-colour lower/light printed fabric.

35

Albatros D.Va

Vzfw. Jupp Cremer, Jasta 5, Lieramont, Winter/Spring 1918

Jasta markings in red and green.
All surfaces in five-colour printed fabric.

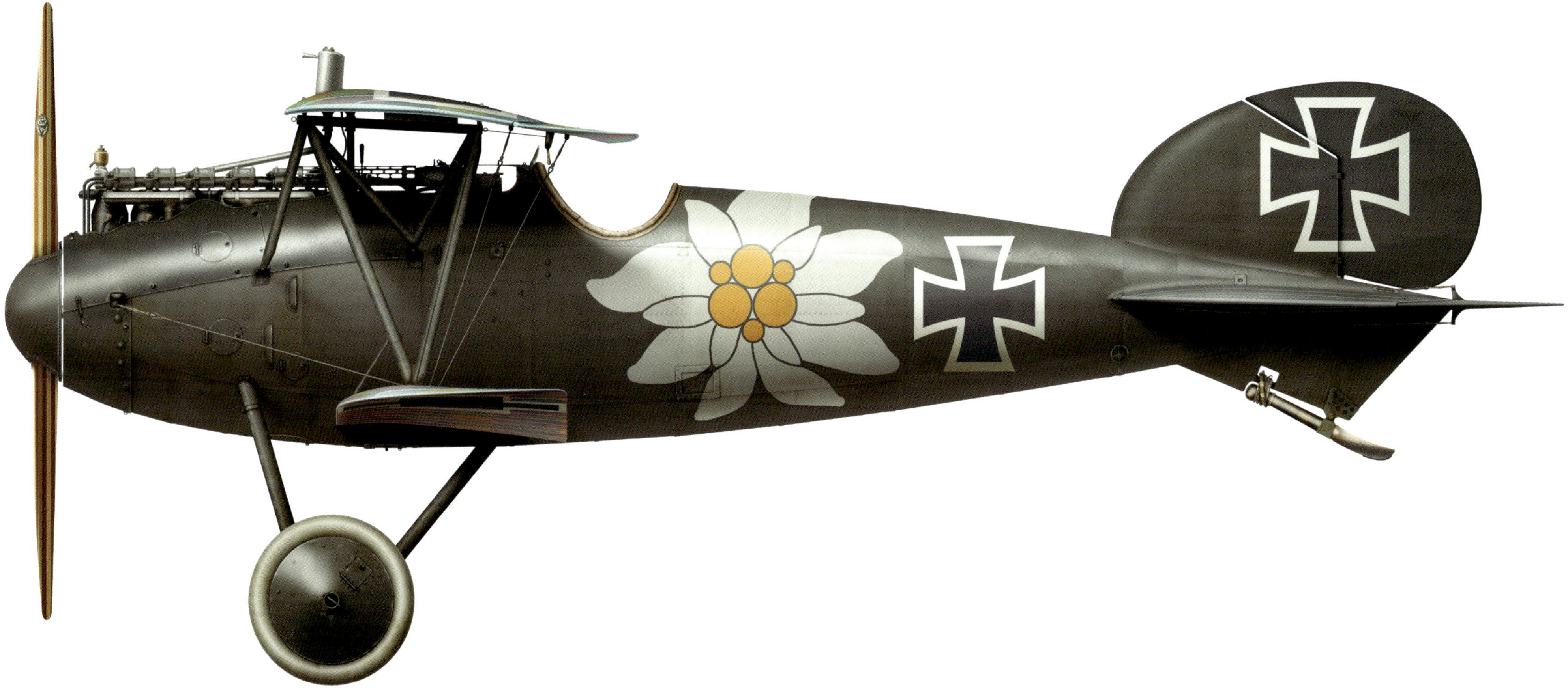

Albatros D.Va

Ltn.d.R Otto Kissenberth, Jasta 23b, Summer 1917

Black fuselage with black and white and yellow personal markings.
Fabric surfaces in five-colour printed fabric.

37

Albatros D.Va (O.A.W.) D.7237/17

Ltn.z.S. Lothar Weiland, Seefrontstaffel 1, July 1918

**Fuselage with green stipple camouflage over varnished plywood. Yellow Jasta markings.
Wings in five-colour printed fabric upper surfaces and light blue lower surfaces.**

Albatros D.Va (O.A.W.) D.6553/17

Jasta 73, May 1918

Black and white Jasta markings over factory finished fuselage.
Fabric surfaces in five-colour printed fabric.

Albatros D.Va (formerly flown by Otto Kissenberth)
Eduard Ritter von Schleich "Der Schwarze Ritter"
Jagdgruppe Nr. 8B, Cambrai-Épinoy, May 1918

Black fuselage. Surfaces in five-colour printed fabric.

Albatros D.Va

Jasta 63, 1918

Black and white Jasta markings over factory finished fuselage.
Wings in five-colour printed fabric upper surfaces and light blue lower surfaces.

Albatros D.Va D.7161/17

Jasta 46w

As it may have appeared in April 1918.
Yellow and green Jasta markings. Fabric surfaces finished in green/mauve upper, light blue lower camouflage.

Selected Bibliography

Ferko, A. E. *Richthofen*. Berkhamsted, Hertfordshire (England): Albatros Productions, 1995
Franks, Norman L. R., and Harry Dempsey. *Albatros Aces of World War I*. Oxford: Osprey, 2000
Gray, Peter L. *The Albatros D.V. Leatherhead*: Profile Publications, 1965
Grosz, P. M. *Albatros D. III, Windsock Datafile Special*. Berkhamsted: Albatros Publications, 2003
Grosz, Peter M. *Albatros D.I/D.II, Windsock Datafile 100*. Berkhamsted, Hertfordshire: Albatros Productions, 2003
Kowalski, Tomasz J. *Albatros D.I-D.Va: Legendary Fighter*. Lublin: Kagero, 2010
Merrill, G. K. *Jagdstaffel 5 Vol. 1 & 2*. Berkhamsted: Albatros Productions, 2004
Mikesh, Robert C. *Albatros D. Va.: German Fighter of World War I*. Washington, D.C.: Published for the National Air and Space Museum by the Smithsonian Institution, 1980
Miller, James F. *Manfred Von Richthofen: The Aircraft, Myths and Accomplishments of 'The Red Baron', FFA69, BAO, K8, J2/B, J11, JG 1*. Crowborough, East Sussex: AirPower Editions, 2009
Miller, James F. *Albatros D.I-D.II*. Oxford: Osprey, 2013
Owers, Colin A. *Albatros D. V/D.* Va at War Vol. 1 & 2, Windsock Datafile 151 & 152. Berkhamsted, Hertfordshire: Albatros Productions, 2012
Rimell, Ray. *Albatros DIII, Windsock Datafile 1*. Berkhamsted: Albatros Productions, 1987
Rimell, Ray. *Albatros D V, Windsock Datafile 3*. Berkhamsted: Albatros Productions, 1987
Rimell, Ray. *Albatros D. II, Windsock Datafile 11*. Berkhamsted: Albatros Productions, 1988
Rimell, Ray. *Albatros Fighters*. Berkhamsted, Hertfordshire: Albatros Productions, 1991
VanWyngarden, Greg. *Von Richthofen's Flying Circus: Colours and Markings of Jagdgeschwader Nr.1*, Windsock Fabric Special No. 1. Berkhamsted, Hertfordshire: Albatros Productions, 1995
VanWyngarden, Greg, and Dempsey, Harry. *'Richthofen's Flying Circus': Jagdgeschwader Nr 1*. Oxford: Osprey, 2004
VanWyngarden, Greg. *Jagdstaffel 2 'Boelcke': (Von Richthofen's Mentor)*. Oxford: Osprey Publ., 2007
VanWyngarden, Greg, and Dempsey, Harry. *Albatros Aces of World War I Part 2*. Oxford, UK: Osprey Pub., 2007
VanWyngarden, Greg, and Dempsey, Harry. *Jasta 18: The Red Noses*. Oxford: Osprey Pub., 2011
VanWyngarden, Greg, and Dempsey, Harry. *Aces of Jagdstaffel 17*. Oxford: Osprey Pub., 2013

VanWyngarden, Greg, and Dempsey, Harry. *'Richthofen's Flying Circus': Jagdgeschwader Nr 1*. Oxford: Osprey, 2004
VanWyngarden, Greg, and Dempsey, Harry. *Albatros Aces of World War I Part 2*. Oxford, UK: Osprey Pub., 2007
VanWyngarden, Greg, and Dempsey, Harry. *Aces of Jagdstaffel 17*. Oxford: Osprey Pub., 2013

Published in Poland in 2014
by STRATUS s.c.
Po. Box 123,
27-600 Sandomierz 1, Poland
e-mail: office@mmpbooks.biz
for
Mushroom Model Publications,
3 Gloucester Close,
Petersfield,
Hampshire GU32 3AX
e-mail: rogerw@mmpbooks.biz

ISBN
978-83-63678-57-9

Editor in chief
Roger Wallsgrove

Editorial Team
Bartłomiej Belcarz
Robert Pęczkowski
Artur Juszczak

Colour profiles
Dave Douglass

DTP
Artur Juszczak

Printed by
Drukarnia Diecezjalna,
ul. Żeromskiego 4,
27-600 Sandomierz
www.wds.pl
marketing@wds.pl

PRINTED IN POLAND

Albatros D.Va (O.A.W.) 6553/17 of Jasta 73. Photo courtesy of Greg VanWyngarden.

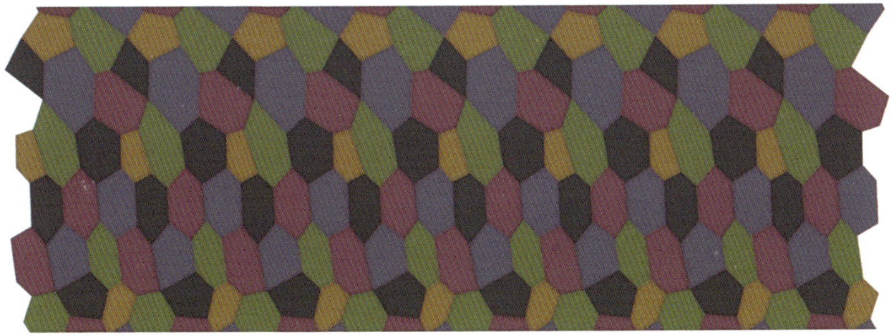

Uppersurfaces, darker printed fabric sample.

Undersurfaces, lighter printed fabric sample.